A History of Computer Operating Systems

Unix, DOS, Lisa, Macintosh, Windows, Linux

By Jon Watson
NIMBLE BOOKS LLC

Jon Watson

ISBN-13: 978-1-934840-45-0

ISBN-10: 1-934840-45-9

Copyright 2008 Jon Watson

Version 1.0; last saved 2008-06-26.

Nimble Books LLC

1521 Martha Avenue

Ann Arbor, MI 48103-5333

http://www.nimblebooks.com

This book was produced using Microsoft Word 2007 and Adobe Acrobat 8.1. The cover was produced using The Gimp. The cover font, heading fonts and the body text inside the book are in Constantia, designed by John Hudson for Microsoft.

Contents (including Figures)

Introduction	v
About the Author	vi
Acknowledgements	vi
The Operating System	1
Figure 1. An AWARD (TM) BIOS chip from 1995	1
Pre-Operating System Computers	3
Mainframes	5
Figure 2. Honeywell-Bull DPS 7 Mainframe BWW March 1990	5
UNIX	6
Figure 3. The UNIX File System Idea	7
Apple's Operating Systems	10
Figure 4. Apple Lisa with a ProFile hard drive stacked on top of it.	10
DOS	14
MS-DOS	14
Figure 5. IBM PC 5150 Running DOS 5.0	16
IBM DOS	17
Apple DOS	18
Other DOSes	19
Microsoft Windows	20
Consumer Time Line	21
Windows Vista	24
Windows Servers	24
Windows Server 2000	25
Windows Server 2003 Family	26
Windows Server 2008 Family	27
Linux	28
Figure 6. Ubuntu Linux With the Gnome Desktop	28
The Naming Controversy	31
Figure 7. The GNU Project Logo	32
Operating Release Timeline	33
The Big List of Operating Systems	35
Acorn	35
Amiga	35
Apollo	35
Apple	36
Atari	36
Burroughs (later Unisys)	36
Convergent Technologies (Later acquired by Unisys.)	37
Be Incorporated	37

Digital/Tandem Computers/Compaq/HP	37
Fujitsu	38
Green Hills Software	38
Hewlett-Packard	38
Intel	38
IBM	38
ICL (formerly ICT)	40
Micrium	41
Microsoft	41
SCO / The SCO Group	42
Lisp and other languages	43
Other	43
Other proprietary UNIX-like and POSIX-compliant	44
Research UNIX-like and other POSIX-compliant	45
Free UNIX-like (aka open source)	46
OpenBSD forked from NetBSD	46
Open source non-UNIX-like	46
Disk Operating System	47
Network Operating Systems	47
Web operating systems	48
Personal digital assistants (PDAs)	48
Music players	49
Smartphones	49
Colophon	50
Figure 8. When a Windows operating system fails: the blue screen of death.	51

INTRODUCTION

The speed at which computers have penetrated almost every industry and household on the planet can be attributed to the breakneck pace at which hardware and software has evolved. While the hardware of any given computer system does the actual work, it is just so much useless junk without an operating system to coordinate it and run applications. The almighty operating system must continue to evolve at frantic pace in order to service humanity's never-ending thirst for machines, gadgets, and entertainment. This book will recount the brief history of the computer operating system, focusing mainly on the UNIX and UNIX-like OSes; DOS; Windows; and Mac.

It's the beginning of the 21st century. Computers in a form that we would recognize today have been around for less than 50 years and they are advancing so fast that we never really understood them to begin with. They went from subservient boxes with blinking lights doing the bidding of academics in the darkest of labs to driving our cars and routing our telephone calls. It's a good thing computers are our friends because we never even got the chance to learn how they work before they took over our lives. How did this meteoric rise take place? Largely by the speedy and creative evolution of the operating system.

ABOUT THE AUTHOR

Jon Watson is an independent GNU/Linux "gun for hire" consultant living in the beautiful maritime province of Nova Scotia, Canada. After graduating with a Computer Information Systems diploma from Mount Royal College in Calgary, Alberta, Jon proceeded to gain both a CompTIA Linux+ certification and an ExpertRating Linux certification while building up his open source consulting business (http://www.jonwatson.ca). Jon has several published articles in mainstream Linux magazines such as Linux Journal and Linux Pro and has spoken at various Linux and open source related events. Jon's current passion is running the Phoenix Hollow Bed and Breakfast (http://www.phoenixhollow.com) with his wife Kelly out of their century home in Windsor, Nova Scotia.

ACKNOWLEDGEMENTS

This one is for my wife, my love, my friend, and my tireless cheerleader; for the one person who thinks I am better than I am. This is for my wife, the indomitable Kelly Mitchelmore.

THE OPERATING SYSTEM

Arguably, the most important part of any computer is its operating system. While a typical computer boot process doesn't call the operating system right away, it hands off control to the operating system before any real work can be done.

When a computer is powered on, a small bit of code usually stored in firmware on the motherboard executes. This code is typically called the Basic Input/Output System (BIOS) and it awakens a computer with just enough knowledge to do three things: activate an input device which is usually a keyboard; activate an output device, which is usually a monitor and, finally, find the operating system kernel, which usually means bringing a disk drive to life so the BIOS can pass control over to the kernel. All of this occurs in the first few seconds of a computer's boot cycle, and everything else from that point on is controlled by the operating system.

Figure 1. An AWARD (TM) BIOS chip from 1995[1]

Today's de facto definition of an operating system generally includes everything that is available when a computer is turned on, including all applications such as word processors, games, and web browsers. This is a common misconception. In fact, the operating system itself is only the part of the machine that manages system resources such as memory and disk

---

[1] http://commons.wikimedia.org/wiki/Image:AWD_BIOS.jpg

drives. In essence, the kernel is the operating system and it is not meant to be used by humans. The applications that an operator uses are just that—applications. They are not part of the operating system proper.

This is a difficult distinction to see because commercial operating system vendors such as Microsoft and Apple recognize that today's typical computer user neither needs to know about the kernel nor would care if they did know. People don't run computers for the sake of running computers. They run them to *do something* and those *somethings* can only be done by applications that sit on top of the operating system.

It's not too far a walk back in time before this separation becomes somewhat apparent. Microsoft Windows 3.0 and 3.1, for example, required Microsoft DOS to be installed first, as the operating system, and then Microsoft Windows could be installed in order to provide a nice graphical user interface. MS-DOS was also much more than an operating system as it came with many applications and utilities, but it contained an actual operating system that the application—MS Windows in this case—could use. In 1995, Microsoft Windows 95 was released with its own kernel, and the visible separation between operating system and applications was erased for Windows users forever.

All other predominant consumer operating systems today have much the same history. They almost all went through a command-line phase in which the operating system and the applications were tightly linked.[2] Those that survive today invariably have evolved to include a graphical user interface so that less technical users can get around.

---

[2] Neal Stephenson's *In the Beginning ... Was the Command Line* (Harper Perennial, 1999)_is a good read on this topic.

PRE-OPERATING SYSTEM COMPUTERS

Early computers did not have operating systems. They were more mechanical than computational in nature and users provided both the application itself and the data to be processed. There was no controlling operating system to ensure that the computer's resources were used properly and the computer could only execute one program at a time. Programs, in this context, were literally just computational tasks that ran from start to finish and produced some output. Today's programs are so complex that they are referred to as *applications* and have no such clearly defined lifespan. By way of contrast, an early program might calculate the length of time it would take to ride a bicycle around the earth while today's applications are things like web browsers and word processors which bear little resemblance to computational tasks.

Diagnostic output from the early machines was largely comprised of blinking lights and occasional puffs of smoke. Debugging a program gone bad on these early machines was a difficult process because if a program crashed, there was no supervising operating system to detect the crash, try to deal with it gracefully, and attempt to provide some sort of output to the developer or user.

The first steps on the road to the modern day operating system were the addition of code libraries to these machines. Developers could then program their applications to make use of these libraries and not only significantly reduce their own code, but also have some assurances that the tasks run by those libraries were being run correctly.

As computers evolved, the time it took to execute programs became less and less, but the time required to switch between programs remained the same. The delta continued to grow until it became obvious that spending 30 minutes to load a program that would take 60 seconds to run wasn't a scalable use of time. In order to combat this, the runtime libraries were expanded and monitor programs were developed which took over some of the grunt work of loading programs into the computer, running the program, calculating executing time usage, and then loading the next program. Computers were starting to be able to look after themselves and this was the first glimmer of the modern day operating system.

MAINFRAMES

The term mainframe refers to a category of computers where operating systems really came into their own. Mainframe computers were the first computers capable of multi-tasking. While initial mainframes operated in batch mode (one job at a time), it wasn't long before the technology evolved to allow multiple jobs to be run at the same time. The name *Mainframe* probably comes from the fact that the first models were extremely large and took up entire rooms filled with metal frames to which the hardware was bolted.[3] So, while mainframes aren't operating systems, no book on the history of operating systems could be complete without discussing their role.

The mainframe era is traditionally thought of as the 1950's to 1970's, but there are many mainframes in use today. Initially, mainframes were "The Next Big Thing" in computing and they allowed multiple users to connect to the mainframe via remote terminals to perform their programming jobs. As computers became more powerful and smaller, many organizations opted out of the massive expense of the traditional mainframe, but in industries where reliability and high-availabilty are mission critical, mainframes still enjoy great use.

Figure 2. Honeywell-Bull DPS 7 Mainframe BWW March 1990[4]

---

[3] http://en.wikipedia.org/wiki/Mainframe_computer
[4] http://en.wikipedia.org/wiki/Image:Honeywell-Bull_DPS_7_Mainframe_BWW_March_1990.jpg

UNIX

These days, there are many applications that run on multiple operating systems, such as the Internet Explorer web browser for Windows and MacOS, and the OpenOffice.org office suite for Windows, and GNU/Linux. This freedom of movement among varying operating systems wasn't available in the early days and is still often a challenge. Not only was it impossible for computers from different vendors to talk to each other in any meaningful way, it was also common for different versions of computers from the same vendor to be incompatible. This quickly became tiresome for business and academia because upgrading or changing computer systems frequently meant re-entering all data and configuration information, which was a time consuming and error prone process.

In 1965, a group of computer organizations including MIT, Bell Labs, and GE tried to create a better solution—an operating system that was usable and interactive named Multics (Multiplexed Information and Computing Service). While Multics did become a functioning operating system, it was more a lab experiment than anything practical and useful in the real world. So although Multics ultimately failed to achieve its initial goals, it formed the foundation for a group of AT&T computer scientists at Bell Labs to write the UNIX operating system in the C programming language[5].

The origins of UNIX can be traced to sometime in early 1969, when engineers at Bell Labs and AT&T started to conceptualize and document how the operating system should be designed and actual hard copy notes became available to developers so they could start work.

The first fledgling thread of what would become UNIX was actually implemented on a PDP-7 when some engineers ported a game named "Space Travel" (written by Bell Labs engineer Ken Thompson for MULTICS). Space Travel was a little too CPU-hungry to be played on the current mainframes, so Thompson and fellow Bell Labs hacker Dennis

---

[5] http://www.bell-labs.com/history/unix/chaos.html

Ritchie rewrote it to run on a PDP-7 and in doing so implemented the file system part of the UNIX plan. UNIX now had a rough plan, and a file system. Things were coming together.

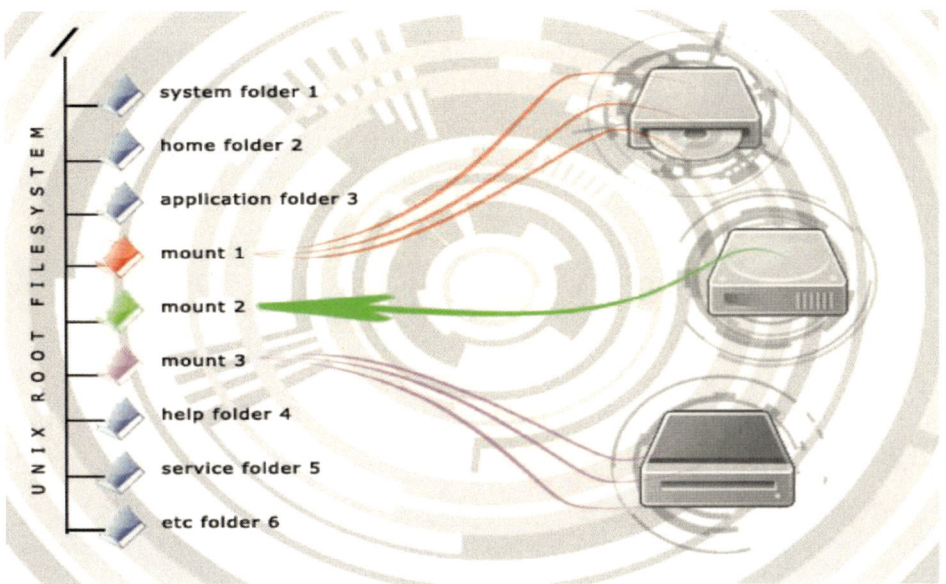

Figure 3. The UNIX File System Idea[6]

At this point, the PDP-7 wasn't yet self-sufficient. Any applications running on it only got there by first being developed using the General Electric Comprehensive Operating System (GECOS) and then transferred to the PDP-7 via tape. By 1970, the group of Bell Labs engineers had developed and implemented file utilities and an assembler. It was that assembler that severed the connection between GECOS and the PDP-7 forever. The PDP-7 became self-sufficient and an operating system proper was born.

Dennis Ritchie said it the best, "Although it was not until well into 1970 that Brian Kernighan suggested the name 'UNIX,' in a somewhat treacherous pun on 'Multics,' the operating system we know today was born."

---

[6] http://commons.wikimedia.org/wiki/Image:Unix_filesystem.png

The stage was set to develop the first practical commercial application for UNIX. The original PDP-7 was rapidly becoming obsolete so Bell Labs purchased a PDP-11 and Thompson and Ritchie et al began porting UNIX to the PDP-11. At the same time, the Bell Labs patent department was actively sourcing an application to assist them in the preparation of patent applications. While developing UNIX to support text processing, it became natural to assist the patent department typists with their work and in the end the patent department adopted UNIX as did many other Bell Labs departments afterwards.

Initial versions of UNIX were written in Assembler, FORTRAN, B, and NB (New B) but by 1972, Ritchie and Thompson started rewriting UNIX in C. There were challenges, but they were eventually overcome. UNIX was now the first operating system to be written in a high level language that could be (relatively) easily ported to other computer systems. That meant that applications written to run on UNIX on one type of computer system could now be run on other computer systems running UNIX. The world was becoming larger.

The next 30 years of UNIX were fast and furious. In the 1970's AT&T proactively distributed UNIX source code and favorable licensing terms to academia and governments so successfully that UNIX became the most taught operating system of the times. Because so many groups and people were working with it and on it, the 70's saw a plethora of different versions of UNIX become available and go into practical use. One of the more well-known versions that is still around today is the Berkeley Software Distribution developed at the University of California, Berkeley (now known as FreeBSD) .[7]

In the 80's, AT&T changed their licensing terms to be less-favorable to academia so many of these groups that were working on UNIX turned their gaze inward and continued to develop their own versions rather than contributing to the whole. Sun Microsystems was founded in 1982 with a version of UNIX that became SunOS and even Microsoft briefly jumped

---

[7] http://en.wikipedia.org/wiki/UNIX

into the fray with a UNIX operating system named XENIX. AT&T still maintained the main stream of UNIX and continued to release versions of their own with UNIX System III and then consolidated into UNIX System V release 1, 2, 3 and 4 by the end of the 80's.

The 1990's were the decade of the "UNIX Wars." Competing groups with different interests formed groups and organizations all over the globe to promote their own standards in the hopes of becoming "the" UNIX. In reality, the UNIX wars ended with a whimper, not a bang, and in general probably did nothing more than scare away the market from UNIX into the waiting arms of Microsoft WindowsNT.[8]

In the 21st Century, the biggest UNIX news has been the ongoing legal battle between Novell and The SCO Group. The background goes something like this: AT&T sold all their rights in UNIX to Novell right after UNIX System V Release 4. Novell then transferred the trademark and certification rights to the newly formed X/open Consortium shortly thereafter. Novell further divested itself of the rights to continue developing UNIX and all existing UNIX licenses to SCO in 1995. In 2000, SCO sold all its UNIX business to Caldera which promptly renamed itself to The SCO Group. The new SCO Group then promptly went to town in an attempt to sue various parties around the globe for using UNIX code in their products as well as violating the UNIX license. In 2007, the bulk of the lawsuit against Novell was settled when the courts decided that Novell did not sell their copyright on UNIX to SCO and therefore, for the most part, SCO had no basis to continue.

The only versions of UNIX that are still doing relatively well today are Solaris, an open source version from SunOS; HP-UX , Hewlett-Packard UNIX; AIX, from IBM; and FreeBSD.

---

[8] http://en.wikipedia.org/wiki/UNIX_wars

Apple's Operating Systems

Apple's history with DOS is documented in the Apple DOS section of this book. This section deals primarily with the period of time from the release of the first Macintosh in 1984 along with the Apple Lisa operating system in 1983 onwards.

The Apple Lisa was the first Apple operating system with a fully graphical user interface. There have been tomes written speculating where the inspiration for this GUI came from as Xerox and Apple made a deal to share some Apple stock in return for the Xerox technology in this area. Xerox's famed Palo Alto Research Center (PARC) had been working on various graphical interfaces and pointing devices for years before the deal with Apple, but Apple had also been working on the Lisa operating system prior as well. There is much speculation on what parts of the operating system were designed in-house and which were inspired by Xerox.

Figure 4. Apple Lisa with a ProFile hard drive stacked on top of it.[9]

In any case, Lisa came with a suite of several office applications and also introduced the world to the basic design of the Apple desktop for years to come. Lisa had a top-menu bar and the proprietary Apple key was

[9] Source: http://en.wikipedia.org/wiki/Image:Apple_Lisa.jpg

present on the keyboard, albeit with an actual apple icon on it rather than the clover-leaf symbol of later years.

Sadly, Lisa cost over $10,000, and therefore really wasn't headed for wide-spread adoption at the consumer level.

Lisa 2 was released in 1984 simultaneously with the Apple Macintosh, but the Macintosh came with its own operating system simply called System 1. System 1 was a single user, single task operating system and at its core was the *Finder*. The Finder is roughly analogous to today's desktop shells in that the user interacted with the system by finding files, opening folders, and the like. However, since System 1 was a single-task operating system, the user had to quit whatever application he or she might have been using before interacting with the finder. Rudimentary by today's standards, but quite intriguing at the time.

Another ground-breaking aspect of System 1 is that it contained a toolbox of code which developers could use to create applications. Today, application programming interfaces (APIs) and software development kits (SDKs) are the de facto ways in which organizations promote rapid development for their systems. In 1984, however, giving developers a library of code to do standard things like build menus and move windows around freed them up to do the real development work of an application's core functionality.

Apple spent the next few years refining System 1 into System Software Releases 2 through 6 with the ultimate reward being the release of System 7 in 1991. Along with the work on the System Software Releases 2 through 6, Apple took a couple of detours and released GS/OS (an operating system designed for the Apple II GS) and A/UX, a version of Apple UNIX.

In 1991, System 7 was released and it heralded what many Macintosh users feel were the greatest advancements in the System since development began. While Systems 7's networking, 32-bit memory implementation and advanced Multi-finder were a great boon to users, it still lacked real memory protection and could only support co-operative multi-tasking.

The next step in Apple's life was to develop the PowerPC. System 7.1.2 was the first System release to support the PowerPC.

The next version of System to be released was System 7.6. This occurred after a somewhat involved interlude during which Steve Jobs left Apple and became a founder of NeXt computing. While this period of time and the development that occurred within the NeXt camp is interesting, it doesn't really fit in with the sequential history of Apple operating systems.

System 7.7 became System 8 before release and things were moving forward once again.

In 1999, System 9 was released and brought with it the ability to be fully updated over the Internet and supported the use of the AppleTalk network protocol over the now-standard TCP/IP. The last release of System 9 was System 9.2.2 in 2001.

MAC OS X hit the world with a whimper as a $29.95 public BETA in September 2000. The actual release of Cheetah (version 10.0) was in March 2001 with version 10.1 (Puma) in September 2001, 10.2 (Jaguar) in August 2002 and 10.3 (Panther) in October of 2004. A fast release schedule which saw a massive amount of lifestyle applications built-in to the system was underway. Existing Mac users were impressed and many PC users converted as Macs were seen to "just work" as compared to the increasingly unstable experience the average Microsoft Windows PCs were perceived to be delivering. Quite possibly the reason for this perceived stability and ease of use was because Mac OS X 10 was based largely on the UNIX variant FreeBSD.

Mac OS X version 4 (Tiger) was released in 2005 and the current version at the time of this writing is Mac OSX 10.5 (Leopard) which was released in October 2007.[10] Leopard will probably be long remembered for its accompanying series of television commercials where a hapless nerd

---

[10] http://en.wikipedia.org/wiki/Mac_OS_X_v10.5

depicting a (Windows) PC and a shaggy twenty-something cool kid representing a Mac talk about how much better Mac is than Windows.⁷

DOS

As computers became more autonomous and capable of self-management, they started to need some non-volatile space to save persistent system files and user data. That need prompted the invention of the disk drive and with that new piece of hardware came the need for the computer to be able to control and manage it. Thus, the Disk Operating System (DOS) was born.

There were (and still are) a number of DOSes out there because as with all new technologies, there was no thought of consistency or cross platform compatibility. In the end, almost all roads lead back to Microsoft DOS, but in the beginning things were a little rough.

MS-DOS

Arguably, the most common and well-known DOS became Microsoft DOS which was developed and marketed as a stand-alone operating system until the advent of Windows95 which saw the amalgamation of the operating system in the graphical operating environment. In reality, DOS version 7.0 still ran under Windows95, but it was well hidden away as Microsoft correctly surmised that end-users are more attracted to a graphical interface with a mouse rather than a hacky, black command line.

As with many of Microsoft's products, MS-DOS did not start out as a Microsoft product. Seattle Computer Products developed QDOS (short for Quick and Dirty Operating System) as the operating system they were waiting for from Digital Research was not coming along fast enough. Microsoft had a single big client then, IBM, and needed a DOS to sell as part of a package. Microsoft licensed the rights to sell QDOS to IBM for $100,000 in 1980.

At that time Seattle Computer Products realized that their Quick and Dirty DOS might actually be a viable long term DOS and changed its name to 86-DOS to reflect the 8086 platform it was designed to run on. Microsoft then licensed non-exclusive rights to sell 86-DOS to any of its clients and finally, in July of 1981 Microsoft bought the complete and exclusive rights for 86-DOS from Seattle Computer Products and adopted the name MS-DOS which became the de facto DOS for the next decade.

Since releasing MS-DOS in 1981, Microsoft has developed and released MS-DOS 1.25 (1982), MS-DOS 2.0 written from scratch (1983), MS-DOS 3 (1984), MS-DOS 4 (1988), MS-DOS 5 (1991), MS-DOS 6 (1993), Windows95 with MS-DOS 7 (1995).[11]

MS-DOS 5.0 arguably signaled the turn in Microsoft's plan of developing and distributing an operating system alone, and developing and distributing a suite of applications to make computers more useful right out of the box. MS-DOS 5.0 contained utilities like undelete and unformat as well as a full screen editor to allow the creation and management of text documents easier.

While MS-DOS 5.0 gave the computing world a taste of what a computer could do with sufficient applications packed into the operating system, MS-DOS 6.0 was heralded by many end-users as one of the most significant releases to date. This was largely because it was packed with so many applications to deal with disks and memory that it really was an operating system and an application suite rather than just an OS. The most memorable function that MS-DOS 6.0 provided was disk compression under the name DoubleSpace. Stac Electronics had a similar product at the time named Stacker and felt that Microsoft's implementation of disk compression violated two Stac patents. Stac Electronics alleged that Microsoft's use of their technology was "willful and deliberate" and "thus Stac [was] further entitled to treble damages, as well as its actual attorneys' fees and litigation costs".[12]

Stac successfully sued Microsoft for the two counts of patent infringement and were awarded $120 million in damages. Microsoft counter-sued that Stac had misappropriated the Microsoft technology that

---

[11] http://members.fortunecity.com/pcmuseum/dos.htm
[12] http://vaxxine.com/lawyers/articles/stac.html

loads the compression drivers in advance of being used which appears in Stacker 3.1. Microsoft was awarded $13.6 million in damages.[13]

Figure 5. IBM PC 5150 Running DOS 5.0[14]

The courts issued an injunction preventing Microsoft from not only continuing to use DoubleSpace, but also from continuing to market it. Microsoft released and distributed MS -DOS 6.21 which lacked any disk compression capabilities which were quickly rewritten and the re-released in MS-DOS 6.22.

---

[13] http://en.wikipedia.org/wiki/DoubleSpace
[14] http://commons.wikimedia.org/wiki/Image:IBM_PC_5150.jpg

IBM DOS

IBM's version of DOS was marketed as PC DOS. Until the early 90's, Microsoft and IBM jointly developed their versions of DOS under an agreement that saw each company working on separate parts of the code. In a very general sense, the agreement saw Microsoft producing the core operating system code and IBM assisted with defining the functionality and testing as well as developing and integrating applications and code that would work specifically with IBM computers.

Both versions of DOS did well but whether that was the outcome of this agreement or the impetus for it is unclear. At the time, there were essentially two classes of personal computers with good market momentum: IBM's computers and non-IBM computers. Since Microsoft was only obligated to provide core code to IBM and not a complete operating system, Microsoft was able to enter into license agreements with many other computer manufacturers whose machines were not 100% IBM compatible and therefore could not run PC DOS out of the box. Companies like Compaq and Texas Instruments licensed copies of MS-DOS and then were able to modify the code to work on their machines and sell them. Eventually, it became obvious that making computers that were not 100% IBM compatible was a bad marketing move and most hardware manufacturers just starting producing IBM compatible machines which fuelled the infamous wave of IBM "clones" in the 1990's that were capable of using IBM's PC DOS rather than purchasing and modifying MS-DOS.[15]

PC DOS came into being in 1980 when Microsoft licensed QDOS from Seattle Computer products in order to sell it to IBM. In 1981 after Microsoft purchased all the rights to DOS from Seattle Computer Products, PC DOS 1.0 was released.

Upgrades to PC DOS were released sporadically with version 2.1 released in 1983, PC DOS 3.3 in 1987, PC DOS 4 in 1988, PC DOS 6.3 in 1994, PC DOS 7 in 1995 (which had Stac Electronics's Stacker disk compression system built-in) and finally PC DOS 2000 was released in

---

[15] http://e-articles.info/e/a/title/MS-DOS-Versus-PC-DOS/

2000 to correct a Y2K problem. PC DOS is still around today in some embedded systems such as IBM ThinkPad's rescue system.[16]

APPLE DOS

Apple DOS versions 1 and 2 were only used internally by Apple and never released to the public. Version 3.0 had enough bugs that it was also never released and the first version released to the consumer market was Apple DOS 3.1 in 1978 to coincide with the Apple II, which had a disk.

In 1979 Apple DOS 3.2 was released to coincide with the release of the Apple II+ and Apple DOS 3.3 was released in 1980. Version 3.3 contained huge gains in disk manipulation and management, but sadly rendered disks formatted in versions prior to Apple DOS 3.3 unreadable. Apple produced a migration utility for these disks named "Muffin" and the user community developed a reverse utility named "Niffum" which could convert Apple DOS 3.3 formatted disk to the legacy format which could be used on earlier Apple DOS machines.

Apple released two more versions of Apple DOS 3.3 in 1983 and then Apple DOS was abandoned and ProDOS was written to replace it. By some accounts, ProDOS was the evolution of the Apple Sophisticated Operating System (SOS) which was released in 1980 in conjunction with the Apple III computer.[17] Apple DOS was very basic in that it could not handle disks with more than 400KB of space on any drive and it was unable to access other disk-like media such as RAM drives, hard drives, or even 3.5-inch floppy disks. While these limitations were not a big deal in the early 80's, as time marched on, it became apparent that Apple DOS would not be able to compete in the DOS market without a complete rewrite.

Rewritten it was and released under the new name ProDOS in 1983. By 1986, 16-bit processors were becoming ubiquitous and ProDOS was forked

---

[16] http://en.wikipedia.org/wiki/PC-DOS
[17] http://www.kernelthread.com/mac/oshistory/1.html

into two versions named ProDOS 8 and ProDOS 16 respectively to support these processors.[18]

By December of 1986 ProDOS was abandoned and the first version of GS/OS hit the market.

OTHER DOSES

There were (and perhaps still are) hundreds of flavors of DOS on the market in the latter decades of the 1900's. While at first glance it may seem like a huge wasted effort for all of companies like Compaq, Texas Instruments, Heath/Zenith, Tandy, et al to invest the time and money to create their own disk operating systems and it certainly would have been, had that been the case.

As explained in the section on IBM PC DOS, most of these manufacturers weren't really creating their own DOSes. Rather, those companies that were marketing non-IBM compatible machines were generally licensing the core code-base of MS-DOS from Microsoft, tweaking it to work on their hardware, producing the manuals and disks required, and then marketing it under their own name. Brilliant in its simplicity.

---

[18] http://apple2history.org/history/ah15.html

MICROSOFT WINDOWS[19]

It's likely that almost every computer user in the world has used the Windows operating environment or system at one time or another. Even die-hard Linux and Mac users probably started their computer lives as Windows users. Regardless of whether you love or hate it, Microsoft Windows has been marketed to perfection and while the precise number of personal computers running Microsoft Windows world-wide is hard to accurately pinpoint, it certainly dominates the market.

---

[19] http://www.microsoft.com/windows/WinHistoryDesktop.mspx

Consumer Time Line

Microsoft Windows made its first appearance as version 1.0 in 1985. It was the first Microsoft operating system to sport a Graphical User Interface (GUI), but it wasn't an operating system proper. Windows 1.0 was essentially a fancy menuing system for the underlying MS-DOS operating system.

Windows 2.0 was released in 1987 and ran on the new and fast 286 processors. One of the great features of Windows 2.0 was that it allowed desktop windows to overlap each other. Windows 2.3 was released to take advantage of the upcoming 386 processors.

Windows 3.0 was really the first version of Windows that made decent headway into the consumer market. Many Windows users today can remember Windows 3.0, but versions before that are becoming lost in the mists of time. Windows 3.0 came out in 1990 and boasted the ability to deal with 16-bit colors, but it still ran on top of the MS-DOS operating system.

Personally, I found the latter half of the 1990s to be a very confusing time for Windows. There were many versions of Windows released during this time and keeping track of what versions were upgrades to which previous version became confusing. From Windows NT to Windows 95 to Windows 2000 to Windows ME, the market was a bit of a confusing mess. With the clarity of hindsight several years later, it is now obvious that there were two main streams of Windows—the networked stream aimed at businesses and the personal stream aimed at end-consumers.

Windows 3.1 was released in 1992 and observers speculate that the release was largely in response to the impending release of IBM's OS/2 operating system. Windows 3.1 had little extra to offer over Windows 3.0 except for some new font technologies and a few bug fixes. The release of Windows 3.1 is curiously absent from the official Microsoft "History of Windows" web page and only blushingly mentioned when used in comparison with Windows NT 3.1

Windows NT 3.1 was released in 1993 and represented a new operating system written completely from scratch. The NT stands for "New Technology" and even though this was the first release of Windows NT, it bore the 3.1 version number to coincide with the already popular Windows 3.1. Under the hood, Windows NT was a 32-bit operating system whereas Windows 3.1 was still running 16-bit code.

The next step in this confusing naming game was the release of Windows for Workgroups 3.11 in 1993. This was an incremental upgrade to Windows 3.1 which essentially just improved LAN networking.

Windows NT Workstation 3.5 was released in 1994 and while it offered a greater degree of data protection for corporate users, its most obvious feature was the introduction of 255-character file names. No longer were users constrained to the old "8 dot 3" naming convention of MS-DOS and burdened with trying to derive intelligent file names.

In 1995 things started coming together a little clearer. Windows 95 was released which had a completely new interface and introduced the world to the "Start button". Rather than having a boxy Program Manager running in the middle of the screen to start a user session, users now clicked on the Start button to gain access to their groups and applications.

Windows 95 was the successor to three Microsoft products: Windows 3.1, Windows for Workgroups, and MS-DOS and was a full-fledged 32-bit operating system.

A notable feature of Windows 95 that was shaky at first, but works well now, is "Plug and Play". Plug and Play was designed to allow users to simply insert new hardware into their computers, such as a new modem or network card, and Windows 95 would simply recognize it and configure it for use. Prior to Plug and Play, each piece of hardware came with drivers which had to be installed in order for Windows to be able to interact with the new device. In order for Plug and Play technology to succeed, hardware vendors had to make their products Plug and Play compatible and it took some time for everyone to come on board and for the existing stock of non-

Plug and Play hardware on the shelves to sell out. In the beginning, computer pundits found Plug and Play to be so unreliable that it was nicknamed "Plug and Pray" in technical circles.

From a networking standpoint, Windows 95 also introduced a 32-bit TCP/IP connection stack to make connecting to the Internet easier and more reliable.

Windows NT Workstation 4 refused to die and was released in 1996. It adopted the Windows 95 interface, but was still aimed at corporate users.

Windows 98 was released in, you guessed it, 1998. During the latter half of the 1990's, Microsoft largely dropped the concept of using version numbers for the products and started using the release year as the version number. Given Microsoft's previous policy of arbitrarily assigning version numbers to their products, this new year-based policy brought some order to the scene. Sadly, they discontinued this policy after 2000 and are now pursuing a policy of completely arbitrary names which are based neither on version numbers, release year, or anything that would indicate anything sequential.

Windows 98 was an upgrade from Windows 95. Amongst its more useful features was the addition of Universal Serial Bus (USB) support. Windows 98 Second Edition (SE) was released shortly thereafter and added many incremental changes as well as support to allow Windows98 SE to use Windows NT hardware drivers.

In the year 2000, Microsoft released Windows 2000 and Windows 2000 Professional. These were the last Microsoft operating systems to be based on the Windows 95 code-base. All future operating systems were to be based on the Windows NT and Windows 2000 kernel.

Windows 2000 was released to replace Windows 95, Windows 98, and Windows NT Workstation 4.

Only a year later, Windows XP and Windows XP Professional were released and the business and personal product lines are merged using the

Windows 2000 code-base. The 'Professional' edition has better file sharing and network functionality and a 64-bit version and the XP Media Centre edition were also released around the same time. The Tablet PC edition was released in 2002.

Windows Vista

Windows Vista was released to the public at large in 2007 and was designed to replace Windows XP. Vista was released in six editions, which although it may seem excessive, can most readily be explained by Microsoft's business need to "segment" its market in the most lucrative way. The editions are Starter, Home Basic, Home Premium, Business, Enterprise, and Ultimate.[17]

The 5-year span between the release of Windows XP and Windows Vista is the longest period of time between any Microsoft releases.

Vista brings a new user interface to the screen which is still largely based on the Windows 95 concept of the Start button, but has a different way of organizing and presenting applications and folders. Microsoft's primary stated goal with Vista was to provide a higher level of security for the computer than was previously attainable using Windows XP.

The user interface is much flashier than previous Windows versions, and the functionality of various parts of the Windows shell is more complex in Vista. Device installation has evolved to the point where Windows Vista will go out on the Internet to find drivers for a piece of hardware it does not recognize and install them automatically.

Windows Servers [18]

Windows NT was really the first Windows server in that it was geared towards corporate networks. The Windows NT product line started with Windows NT 'Advanced Server' 3.1 which was released in 1993 and

---

[17] http://en.wikipedia.org/wiki/Windows_Vista_editions
[18] http://www.microsoft.com/windows/WinHistoryServer.mspx

contained 6 releases over a 5 year period until being discontinued in 1998 in preparation for the Windows 2000 server family release.

In general terms, the functionality of the Windows NT product line followed an inverse relationship with competing server products. Windows NT 3.1, for example, was an application server that was developed to integrate and work with the other popular networking systems of the time such as Novell NetWare and Banyan vines.

In 1994, Windows NT Server 3.5 was released and contained even better tools for integrating with Novell and UNIX environments. However, by the time Windows NT Server 4.0 was released in 1996, the functionality focus was on Microsoft products and integration such as Microsoft's web server, the Internet Information Server. Even the old interface was dropped, and Windows NT Server 4.0 sported the new (at the time) Windows 95 graphical user interface.

The last Windows NT release was the Terminal Server edition in 1998. Terminal server technologies are still in wide use today as they provide remote access to computers by other computers that may or may not be running the same operating system.

WINDOWS SERVER 2000

In 2000, the Windows 2000 Server family was released to work with the Windows 2000 Professional and Windows XP Professional clients. The term 'family' was used for the Server 2000 product line because no less than three editions were made available:

Windows 2000 Server: as it sounds, this is the base server product that was aimed at businesses who needed a basic file and print server, but didn't have the budget or technical requirements to justify a more robust solution.

Windows 2000 Advanced Server was geared toward businesses that required a web presence, and Windows 2000 Datacenter Server was the most robust of the three and had the best scalability and potential for availability.

Windows Server 2003 Family

The Windows 2000 Server family was replaced by the Windows 2003 Server family in 2003 which has no less than eight editions: The SBS (Small Business Server), Web Edition, Standard Edition, Enterprise Edition, Datacenter Edition, Compute Cluster Server Edition, Storage Server, and Home Server.[19]

While the explanation of the capabilities of all of these editions could be the subject of its own book, in a nutshell, here they are:

Windows Server 2003, Small Business Server: In practice, this is one of the more common Windows servers in small to medium sized businesses around North America. It provides typical file, print, and fax server functionality, as well as remote access and remote management of the network and the devices on it.

Windows Server 2003, Web Edition: Primarily intended to be used as a web server and to deal mostly with XML data.

Windows Server 2003, Standard Edition: The Standard Edition is the upgrade from the Small Business Server. It offers more functionality than the Small Business Server and can handle better hardware such as multiple processors, and there is a 64-bit version available.

Windows Server 2003, Enterprise Edition builds on the Standard edition and is aimed at larger businesses. It can handle up to eight processors and more advanced memory access methods as well as clustering.

Windows Server 2003, Datacenter edition is really the head of the family. With support for up to 32, 32-bit or 64-bit processors and the ability to address up to 2TB of RAM, it's quite powerful by anyone's standards.

Curiously, Windows Compute Cluster Server 2003 was released in 2006 but is still considered part of the Windows Server 2003 family. As its name implies, the Windows Compute Cluster edition is meant to be deployed on multiple computers in order to provide supercomputing capabilities.

---

[19] http://en.wikipedia.org/wiki/Windows_Server_2003

The Windows Server 2003, Storage Server is geared toward providing advanced and varied network storage facilities. It has the capability to work with Storage Area Networks and Network Attached Storage devices as well as advanced RAID configurations. It can also provide general file and print sharing services to network users and was not available for retail sale. Customers could only get the Storage Server edition by purchasing it through an Original Equipment Manufacturer (OEM).

Lastly, the Windows Server 2003, Home Edition was released in 2007. This edition was aimed at the growing number of home users who have networks in their house to support printer and file sharing amongst multiple family computers. Aside from file and print sharing, Home Edition also offers some backup capabilities and remote access functions.

WINDOWS SERVER 2008 FAMILY[20]

Like the Windows Server 2003 family, 2008 is comprised of many editions. Nine, in fact—one more than the 2003 group. Released in 2008, many of the same editions exist in 2008 as in 2003, with general upgrades to not only the applications and kernel itself, but also the hardware support. Most editions of the Server 2008 family have 32-bit and 64-bit versions, but some of the higher-end editions such as the Small Business Server and Essential Business Server only come in 64-bit flavors. Microsoft has announced that 32-bit editions in the 2008 family will be the last and all versions henceforth will be 64-bit from this point on.

---

[20] http://en.wikipedia.org/wiki/Windows_Server_2008

Linux

GNU/Linux (or just "Linux" – more on that in the Naming Controversy section) is an open source operating system developed by thousands of individual programmers, governments, and companies world-wide. As such, there is no "Linux, Inc" driving the marketing and release of the operating system so it's not really possible to chronicle the history of Linux in a sequential manner as it is with Microsoft or Apple operating systems.

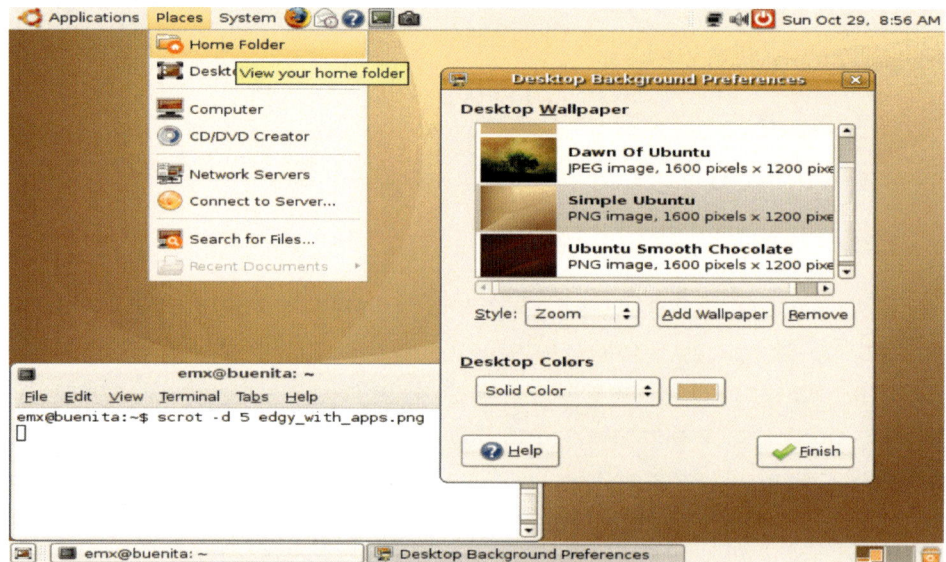

Figure 6. Ubuntu Linux With the Gnome Desktop[21]

Linux is released by these various groups in packages called 'distributions'. There are literally hundreds if not thousands of Linux distributions ranging from the well-known ones such as Red Hat and Ubuntu down to the lesser-known distributions like Musix and Frugalware. Check out DistroWatch.com for a view of how many distributions there are and how popular each one is.

It's fairly easy to chronologically follow the releases of the Linux kernel itself or the history of any one specific distribution, but as a whole, the operating system isn't developed, marketed, or released from a central source. At the time of this writing Red Hat Enterprise Linux Server is at

[21] Source: http://commons.wikimedia.org/wiki/Image:Edgy_with_apps.png

version 5, Fedora Core Linux is at version 9, Ubuntu Linux is at version 8.04, Slackware Linux is at version 12.1, and so on.

The development of the Linux kernel and subsequent addition of other useful tools to become the current day operating environment provides an exceptional instructional opportunity in the separation of the kernel and applications. As discussed briefly in the section on Microsoft Windows and MS-DOS, the actual operating system and the applications that run on it are two different beasts. Taken down one level further, and the actual operating system and the kernel are again two different pieces of the puzzle. Nowhere is this more easily understood than in a cursory look at the history of the Linux kernel and the operating environments that have been build on it.

An operating system is comprised generally of two parts: the kernel, which is responsible for talking to the hardware of the computer, and the "userland" utilities so named because these are the utilities that end-users use. There's a third smudgy line between where "utilities" end and "applications" begin which is also the subject of some debate. While all programs that run on computers can be referred to as 'applications', 'utilities' are generally the type of low-level applications such as text editors and compilers that a developer would use to create higher level applications. Utilities are also generally thought of as applications that are required in order for a computer to be at all useful, such as disk formatters, rather than web-browsers which are not required in order to run the thing.

So, we have some hardware, and then we have a kernel running on top of the hardware so that utilities and applications can make use of it. Then we have some utilities to allow developers to create and maintain useful applications that less-technical, so-called 'end-users' can interact with. Clear as mud? Good, now on with the story.

The Linux kernel was developed by Linus Torvalds in 1991. At the time, Linus was a second-year computer science student at the University of Helsinki. One of his instructors was Andrew Tannenbaum who had written a small operating system named Minix as a teaching aid for his students. At this period in time, there were few operating systems that appealed to hackers as most other operating systems were either closed source or prohibitively expensive. Minix was somewhat popular, but was licensed rather rigidly, which prevented Linus from extending it. So, what would any young hacker do when he needs an operating system and there isn't a suitable one lying around? He'd write one.[22]

August 25th 1991 is considered the birthday of the Linux kernel. That's the date that Linus posted to the Usenet newsgroup comp.os.minix that he was developing a free operating system based on Minix. Andrew Tanenbaum, the creator of Minix, touted Linus' work as 'old-fashioned' and 'obsolete' and predicted that Linux would die a quick death. Clearly, that wasn't the case and that is mostly due to Linus' action of allowing distribution and modification of the Linux source code. In a world of proprietary and closed-source operating systems, such a thing was unheard of, but many hands make light work and the Linux kernel developed by leaps and bounds with a stable version 1.0 being released in March of 1994.[23]

Here's where things get interesting. A kernel is certainly a required part of an operating system because a computer that cannot access its memory or disks is really quite useless. However, a kernel doesn't provide anything directly to the end-user so computer operators sitting in front of a computer with just a kernel running on it are also quite useless. In order for a computer to be truly useful, it needs both a kernel and an operating system.

In 1983, eight years before Linus started work on the Linux kernel, Richard M Stallman founded the GNU project. The GNU project's goal was to create a completely free operating system totally unencumbered by proprietary licensing. Stallman's pedigree could not possibly be taken into issue as his career began in the almost legendary MIT Artificial Intelligence labs, and he was the right man to spearhead this initiative. After watching many of his peers be seduced away by commercial companies and placed under strict non-disclosure agreements for the work they were developing,

---

[22] https://netfiles.uiuc.edu/rhasan/linux/
[23] http://en.wikipedia.org/wiki/History_of_Linux

Stallman pioneered the concept of "copyleft" (vice "copyright") which encapsulates the concept of free software. Stallman, like many other developers and computer users in the world, feel that software should be free for all to use and learn from and he founded The Free Software Foundation in 1985 to promote this goal.

Fast forward to 1994 and we have a Linux kernel with no operating system and a GNU operating system with no kernel. The match wasn't quite made in heaven, but since both Linus' kernel and the GNU project's utilities were all licensed under the GNU General Public License, hackers on both projects had been busily integrating the two parts together for years and an operating system was born.[14]

It should be noted that the GNU project did not just create a bunch of utilities without regard for the fact that they would one day need a kernel. In fact, the GNU kernel, named the HURD, was being developed alongside the utilities. To date, however, the HURD is still not suitable for production use.[24]

The current Linux kernel is up in the 2.6 version branch and has hardware support for an extremely broad and ever-increasing range of hardware. While large networks such as the Internet are still primarily powered by Linux and other UNIX-like operating systems, groups like the KDE and Gnome projects have made great strides in producing mature and stable graphical user interfaces for Linux computers which has promoted Linux as suitable for home users as well. The vast majority of Linux distributions are available free of charge and can be downloaded from various places on the Internet.

THE NAMING CONTROVERSY

While this topic has almost nothing to do with the actual history of the Linux kernel or subsequent distribution releases, it's not possible to talk about Linux and the GNU project without acknowledging the 800lb gorilla in the room.

---

[24] http://www.gnu.org/software/hurd/hurd.html

Some versions of Linux and some people in the Linux community refer to the whole operating systems as "GNU/Linux" (pronounced properly with a hard "G" as in 'Gnew'). Indeed, Stallman and the Free Software Foundation have been advocating the use of the name GNU/Linux as a way to give a nod to both the Linux kernel developers and the GNU developers without both of which the modern day Linux distributions would look much different.

There are two sides to every argument, of course, and proponents of using simply "Linux" to describe the entire operating system feel that listing only GNU as a contributor to the operating system does a disservice to the literally thousands of other contributors over the years . Conversely, including all of them in the name isn't workable. Proponents of the GNU/Linux name feel that since the GNU project supplied the most important pieces of the system (the mid-level utilities that allowed other applications to be built at all) that it is equally as important as the kernel.

It appears, though, that the straight "Linux" camp is winning as there are fewer and fewer uses of "GNU/Linux" as time goes on.

Figure 7. The GNU Project Logo[25]

---

[25] Source: http://www.gnu.org/graphics/agnuhead.html

OPERATING RELEASE TIMELINE

Year	Event
1965	Bell Labs, MIT, GE form a group to build the MULTICS operating system
1969	UNIX starts to take form at Bell Labs
1972	UNIX rewrite in the C programming language begins
1978	Apple DOS 3.1 released
1979	Apple DOS 3.2 released
1980	Microsoft licenses QDOS from Seattle Computer Products for sale to IBM Apple DOS 3.3 released Apple SOS released
1981	Microsoft buys complete rights to QDOS from SCP and renamed it MS-DOS IBM's PC-DOS 1.0 released
1982	Sun Microsystems founded with a UNIX that became SunOS MS-DOS 1.25 released
1983	MS-DOS 2.0 released Apple LISA released PC-DOS 2.1 Richard M Stallman founds the GNU Project
1984	Apple Lisa 2 released MS-DOS 3 released Apple System Software 1 released
1985	The Free Software Foundation is formed Apple System Software 2 released Microsoft Windows 1.0 released
1986	Apple System Software 3 released Apple ProDOS forked into 8 and 16-bit versions: ProDOS 8 and ProDOS 16 released Apple GS/OS released
1987	Apple System Software 4 released PC-DOS 3.3 released Microsoft Windows 2.0 released
1988	MS-DOS 4 released Apple System 6 released PC-DOS 4 released
1990	Microsoft Windows 3.0 released
1991	Apple System 7 released MS-DOS 5 released Linux Torvalds begins work on what would become the Linux kernel
1992	Microsoft Windows 3.1 released

1993	MS-DOS 6 released Microsoft Windows NT 3.1 Microsoft Windows for Workgroups 3.11 released
1994	PC-DOS 6.3 released Version 1.0 of the Linux kernel released Microsoft Windows NT 3.5 released
1995	Windows 95 released with MS-DOS 7 PC-DOS 7 released
1996	Microsoft Windows NT Workstation 4.0 released
1998	Microsoft Windows 98 released Microsoft Windows NT Terminal Server released
1999	Apple System 9 released
2000	PC-DOS 2000 released Microsoft Windows 2000 released
2001	Microsoft Windows XP released Apple MacOS X released Apple MacOS 10.1 released
2002	Apple MacOS 10.2 released
2003	Microsoft Windows Server 2003 family released through 2007
2004	Apple MacOS 10.3 released
2005	Apple MacOS 10.4 released
2007	Apple MacOS 10.5 released Novell wins major part of SCO lawsuit Microsoft Windows Vista family released
2008	Microsoft Windows Server 2008 family released

THE BIG LIST OF OPERATING SYSTEMS

When computing was new, its course was not clear. Largely used only in academia and government, the arrival of a home computer all across the developed world was unforeseen. As computer scientists and engineers attempted to map and develop the course of computer operating systems, a plethora of these systems were created. Many have dropped by the wayside in favour of different operating systems, but a surprising range of operating systems are still in use today.

The following list does not claim to be definitive, but it does present a large portion of the more mainstream operating systems that have paraded across the short history of computing.

Here, then, I present you with The Big List of Operating Systems:[26]

ACORN

- Arthur
- ARX
- OS (on the BBC Micro and BBC Master)
- RISC OS
- RISC iX (based on 4.3BSD)

AMIGA

- AmigaOS
- Amiga UNIX, a.k.a. Amix

APOLLO

- AEGIS/Domain/OS One of the first network-based systems. Ran on Apollo/Domain hardware. Later bought by Hewlett-Packard.

---

[26] http://en.wikipedia.org/wiki/List_of_operating_systems

Apple

- Apple DOS
- ProDOS
- GS/OS
- SOS (Sophisticated Operating System)
- Lisa OS
- System Software 1 through 7
- Mac OS 8
- Mac OS 9
- A/UX
- MkLinux
- Mac OS X v10.0 (aka Mac OS X 10.0 "Cheetah")
- Mac OS X v10.1 (aka Mac OS X 10.1 "Puma")
- Mac OS X v10.2 (aka Mac OS X 10.2 "Jaguar")
- Mac OS X v10.3 (aka Mac OS X 10.3 "Panther")
- Mac OS X v10.4 (aka Mac OS X 10.4 "Tiger")
- Mac OS X v10.5 (aka Mac OS X 10.5 "Leopard")
- Mac OS X Server
- Darwin (open source underpinnings of Mac OS X, based on FreeBSD and NextStep)

Atari

- Atari DOS (for 8-bit computers)
- Atari TOS
- Atari MultiTOS

Burroughs (later Unisys)

- BTOS
- MCP (Burroughs Large Systems)

Convergent Technologies (Later acquired by Unisys.)

- CTOS

Be Incorporated

- BeOS
- BeIA
- ZETA

Digital/Tandem Computers/Compaq/HP

- OS/8
- ITS (for the PDP-6 and PDP-10)
- Multi-Programming Executive (from HP)
- TOPS-10 (for the PDP-10)
- WAITS (for the PDP-6 and PDP-10)
- TENEX (from BBN, for the PDP-10)
- TOPS-20 (for the PDP-10)
- RSTS/E (multi-user time-sharing OS for PDP-11s)
- RSX-11 (multiuser, multitasking OS for PDP-11s)
- RT-11 (single user OS for PDP-11)
- VMS (originally by DEC, now by HP) for the VAX mini-computer range, Alpha and Intel Itanium 2; later renamed OpenVMS)
- Domain/OS (originally Aegis, from Apollo Computer who were bought by HP)
- RTE HP's Real Time Executive (ran on the HP 1000)
- TSB HP's Time Share Basic (yes, it was an operating system, ran on the HP 2000 series)
- Digital UNIX (derived from OSF/1, became HP's Tru64 UNIX)
- HP-UX
- Ultrix
- Guardian
- OSS (POSIX-compliant Open System Services)

Fujitsu

- Towns OS

Green Hills Software

- INTEGRITY Reliable Operating system
- INTEGRITY-178B A DO-178B certified version of INTEGRITY.
- µ-velOSity A lightweight microkernel.

Hewlett-Packard

- MPE Multi-programming Executive; ran on HP3000 mini-computers.
- HP-UX HP-UX; runs on HP9000 and Itanium servers—from small to mainframe-class computers.

Intel

- iRMX real-time operating system originally created to support the Intel 8080 and 8086 processor families in embedded applications

IBM

- IBM 7090/94 IBSYS
- SYSTEM 1400/1800 IJMON A Bootable serial I/O monitor for loading programs.
- BOS/360 Early interim version of DOS/360, briefly available at a few Alpha & Beta System 360 sites.
- TOS/360 Similar to BOS above and more fleeting, able to boot and run from 2x00 series tape drives.
- DOS/360 Disk Operating System. First commonly available OS for System/360 due to problems in the OS/360 Project. Multi-programming system with up to 3 partitions.
- DOS/360/RJE DOS/360 with a control program extension that provided for the monitoring of Remote Job Entry hardware (Card Reader & Printer) connected by dedicated phone lines.
- DOS/VSE First DOS offered on System/370 systems, provided Virtual Storage Extensions, and SNA. Still had fixed size processing partitions, but up to 14 partitions.

- DOS/VSE/ESA DOS/VSE extended virtual memory support to 32 bit addresses (Extended System Architecture).
- z/VSE Latest version of the four decades old DOS lineage. Now supports 64 bit addresses, Multiprocessing, Multiprogramming, SNA, TCP/IP, and some virtual machine features in support of Linux workloads. (All DOS ref. IBM website)
- OS/360 First official OS targeted for the System/360 architecture, saw customer installations of the following variations:
- PCP Primary Control Program, a kernel and a ground breaking automatic space allocating file system.
- MFT Multi-Programming Fixed Tasks, had 15 fixed size partitions defined at boot time.
- MVT Multi-Programming Variable Tasks, had up to 15 partitions defined dynamically.
- RTOS Real Time Operating System, run on 5 NASA custom System/360/75s. A mash up by the Federal Systems Division of the MFT system management, PCP basic kernel and file system, with MVT task management and FSD custom real time kernel extensions and error management. The pinnacle of OS/360 development.
- OS/370 The official port of OS/360 targeted for the System/370 virtual memory architecture. Customer installations in the following variations:
- OS/VS1 Virtual-memory version of OS/MFT
- OS/VS2 Virtual-memory version of OS/MVT
- SVS Single Virtual Storage (both VS1 & VS2 began as SVS systems)
- MVS Multiple Virtual Storage (eliminated any need for VS1)
- OS/390 Upgrade to MVS, with an additional UNIX-like environment.
- z/OS z/Architecture version of OS/390.
- TPF z/OS extension
- CP/CMS Control Program / Cambridge Monitor System, Virtual Machine operating System for System/360 Model 44 and 67
- VM/CMS Virtual Machine / Conversational Monitor System, VM (operating system) for System/370 with Virtual Memory.

- VM/XA VM (operating system) eXtended Architecture for System/370 with extended Virtual Memory.
- VM/ESA Virtual Machine /Extended System Architecture, added 32 bit addressing to VM series.
- z/VM z/Architecture version of the VM OS (64 bit addressing).
- IBM System/34, 36 System Support Program, or SSP
- OS/400 descendant of System/38 CPF
- i5/OS extends OS/400 with significant interoperability features.
- AIX (a System V UNIX version)
- AOS (a BSD UNIX version)
- GNU/Linux (IBM has contributed much code to this open source operating system, listed below)
- PC-DOS IBM supported, documented, and licensed copies of Microsoft MS-DOS
- OS/2 (developed jointly with Microsoft)
- OS/2 Warp
- eComStation (Warp 4.5/Workspace on Demand, rebundled by Serenity Systems International)
- IBM 8100 DPCX
- IBM 8100 DPPX
- K42 PowerPC or Intel x86 based cache-coherent multiprocessor systems (IBM Website)
- IBM EDX Event Driven Executive for the IBM/Series 1 minicomputers
- IBM RPS Realtime Programming System for the IBM/Series 1 minicomputers

ICL (formerly ICT)

- GEORGE 2/3/4 GEneral ORGanisational Environment, used by ICL 1900 series mainframes
- VME by International Computers Limited (ICL), particularly appearing on the ICL 2900 Series

MICRIUM

- MicroC/OS-II (Small pre-emptive priority based multi-tasking kernel)

MICROSOFT

- Xenix (licensed version of UNIX; licensed to SCO in 1987)
- MSX-DOS (developed by MS Japan for the MSX 8-bit computer)
- MS-DOS (developed jointly with IBM, versions 1.0–6.22)
- Windows CE (OS for handhelds, embedded devices, and real-time applications that is similar to other versions of Windows)
- Windows CE 3.0
- Windows Mobile (based on Windows CE, but for a smaller form factor)
- Windows CE 5.0
- DOS based Windows
- Windows 1.0
- Windows 2.0
- Windows 3.0 (the first version to make substantial commercial impact)
- Windows 3.1x
- Windows 3.2 (Chinese-only release)
- Windows 95 (aka Windows 4.0)
- Windows 98 (aka Windows 4.1)
- Windows Millennium Edition (often shortened to Windows Me) (aka Windows 4.9)
- OS/2 (developed jointly with IBM)
- Windows NT
- Windows NT 3.1
- Windows NT 3.5
- Windows NT 3.51
- Windows NT 4.0
- Windows 2000 (aka Windows NT 5.0)

- Windows XP (aka Windows NT 5.1) (codename: Whistler)
- Windows Server 2003 (aka Windows NT 5.2) (codename: Whistler Server)
- Windows Fundamentals for Legacy PCs (aka Windows NT 5.1)
- Windows Vista (aka Windows NT 6.0) (codename: Longhorn)
- Windows Server 2008 (aka Windows NT 6.0) (codename: Longhorn Server)
- Windows 7 (previously codenamed Blackcomb, then Vienna)
- Windows Preinstallation Environment (WinPE)
- Singularity - A research operating system written mostly in managed code (C#)

SCO / THE SCO GROUP

- Xenix, UNIX System III based distribution for the Intel 8086/8088 architecture
- Xenix 286, UNIX System V Release 2 based distribution for the Intel 80286 architecture
- Xenix 386, UNIX System V Release 2 based distribution for the Intel 80386 architecture
- SCO UNIX, SCO UNIX System V/386 was the first volume commercial product licensed by AT&T to use the UNIX System trademark (1989). Derived from AT&T System V Release 3.2 with an infusion of Xenix device drivers and utilities plus most of the SVR4 features
- SCO Open Desktop, the first 32-bit graphical user interface for UNIX Systems running on Intel processor-based computers. Based on SCO UNIX
- SCO OpenServer 5, AT&T UNIX System V Release 3 based
- UNIXWare 2.x, based on AT&T System V Release 4.2MP
- UNIXWare 7, UNIXWare 2 kernel plus parts of 3.2v5 (UNIXWare 2 + OpenServer 5 = UNIXWare 7). Referred to by SCO as SVR5
- SCO OpenServer 6, SVR5 (UNIXWare 7) based kernel with SCO OpenServer 5 application and binary compatibility, system administration, and user environments

Unicoi Systems

- Fusion RTOS highly prolific, license free Real-time operating system.
- DSPOS was the original project which would become the royalty free Fusion RTOS.

LISP AND OTHER LANGUAGES

- LISP machine Operating Systems ran on specialized processors with LISP implemented as microcode
- Symbolics Genera written in a systems dialect of the Lisp programming language called ZetaLisp. Genera was ported to a virtual machine for the DEC Alpha line of computers.
- Texas Instruments' Explorer Lisp machine workstations also had systems code written in Lisp Machine Lisp.
- The Xerox 1100 series of Lisp machines ran an operating system that was also ported to virtual machine called "Medley."
- Lisp Machines, Inc. also known as LMI, also ran an operating system based on MIT's ZetaLisp.
- The Mesa programming language was used to implement the Pilot operating system, used in Xerox Star workstations.
- PERQ Operating System (POS) was written in PERQ Pascal.

OTHER

- EOS (Operating System), developed by ETA Systems for use in their ETA-10 line of supercomputers
- EMBOS, developed by Elxsi for use on their mini-supercomputers
- GCOS is a proprietary Operating System originally developed by General Electric
- PC-MOS/386—DOS-like, but multiuser/multitasking
- SINTRAN III—an operating system used with Norsk Data computers.
- THEOS
- TinyOS
- TRS-DOS A floppy-disk-oriented OS supplied by Tandy/Radio Shack for their Z80-based line of personal computers.

- TX990/TXDS, DX10 and DNOS—proprietary operating systems for TI-990 minicomputers
- MAI Basic Four—An OS implementing Business Basic from MAI Systems.
- Michigan Terminal System—Developed by a group of American universities for IBM 360 series mainframes
- MUSIC/SP (an operating system developed for the S/370, running normally under VM)
- TSX-32, a 32-bit operating system for x86 platform.
- OS ES An operating system for ES EVM
- Prolog-Dispatcher—used to control Soviet Buran space ship.

OTHER PROPRIETARY UNIX-LIKE AND POSIX-COMPLIANT

- Aegis (Apollo Computer)
- Amiga UNIX (Amiga ports of UNIX System V release 3.2 with Amiga A2500UX and SVR4 with Amiga A3000UX. Started in 1989, last version was in 1992)
- Clix (Intergraph's System V implementation)
- Coherent (UNIX-like OS from Mark Williams Co. for PC class computers)
- DNIX from DIAB
- DSPnano RTOS (POSIX nanokernel, DSP Optimized, Open Source)
- Idris workalike from Whitesmiths
- INTERACTIVE UNIX (a port of the UNIX System V operating system for Intel x86 by INTERACTIVE Systems Corporation)
- IRIX from SGI
- MeikOS
- NeXTSTEP (developed by NeXT; a UNIX-based OS based on the Mach microkernel)
- OS-9 UNIX-like RTOS. (OS from Microware for Motorola 6809 based microcomputers)
- OSF/1 (developed into a commercial offering by Digital Equipment Corporation)
- OPENSTEP

- QNX (POSIX, microkernel OS; usually a real time embedded OS)
- Pardus (Turkish Linux)
- Rhapsody (an early form of Mac OS X)
- RISC/os (a port by MIPS of 4.3BSD to the RISC MIPS architecture)
- RMX
- SCO UNIX (from SCO, bought by Caldera who renamed themselves SCO Group)
- SINIX (a port by SNI of UNIX to the RISC MIPS architecture)
- Solaris (Sun's System V-based replacement for SunOS)
- SunOS (BSD-based UNIX system used on early Sun hardware)
- SUPER-UX (a port of System V Release 4.2MP with features adopted from BSD and Linux for NEC SX architecture supercomputers)
- System V (a release of AT&T UNIX, 'SVR4' was the 4th minor release)
- System V/AT, 386 (The first version of AT&T System V UNIX on the IBM 286 and 386 PCs, ported and sold by Microport)
- Trusted Solaris (Solaris with kernel and other enhancements to support multilevel security)
- UniFlex (UNIX emulating OS by TSC for DMA-capable, extended addresses, Motorola 6809 based computers; eg SWTPC, GIMIX, ...)
- Unicos (the version of UNIX designed for Cray Supercomputers, mainly geared to vector calculations)
- Unison RTOS (Multicore RTOS with DSP Optimization)
- DG/UX (Data General Corp)

RESEARCH UNIX-LIKE AND OTHER POSIX-COMPLIANT

- Minix (study OS developed by Andrew S. Tanenbaum in the Netherlands)
- Plan 9 (distributed OS developed at Bell Labs, based on original UNIX design principles yet functionally different and going much further)
- Inferno (distributed OS derived from Plan 9, originally from Bell Labs)
- Plan B (distributed OS derived from Plan 9 and Off++ microkernel)

- Solaris, contains original UNIX (SVR4) code (code now open source via OpenSolaris project)
- UNIX (OS developed at Bell Labs ca 1970 initially by Ken Thompson)
- Xinu, (Study OS developed by Douglas E. Comer in the USA)

FREE UNIX-LIKE (AKA OPEN SOURCE)

- BSD (Berkeley Software Distribution, a variant of UNIX for DEC VAX hardware)
- FreeBSD (one of the outgrowths of UC Regents' abandonment of CSRG's 'BSD UNIX')
- DesktopBSD FreeBSD distribution for desktop use
- PC-BSD FreeBSD distribution for desktop use
- DragonFly BSD forked from FreeBSD
- NetBSD (one of the outgrowths of UC Regents' abandonment of CSRG's 'BSD UNIX')

OPENBSD FORKED FROM NETBSD

- GNU
- Linux
- OpenDarwin
- OpenSolaris, contains original UNIX (SVR4) code
- SSS-PC Developed at Tokyo University
- Syllable
- VSTa
- UNIXLite

OPEN SOURCE NON-UNIX-LIKE

- FullPliant (programming language based)
- FreeDOS (open source DOS variant)
- FreeVMS (open source VMS variant)
- Haiku (open source inspired by BeOS, under development)

- ReactOS (free software Windows NT compatible OS, in early development)
- osFree (open source OS/2 implementation)

DISK OPERATING SYSTEM

- DR-DOS (Digital Research's [later Novell, Caldera, ...] DOS variant)
- Concurrent DOS (Digital Research's first multiuser DOS variant)
- Multiuser DOS (Digital Research's [later CCI's. Real's/...] multiuser DOS variant)
- FreeDOS (open source DOS variant)
- ProDOS (operating system for the Apple II series computers)
- PTS-DOS (DOS variant by Russian company Phystechsoft)
- 86-DOS (developed at Seattle Computer Products by Tim Paterson for the new Intel 808x CPUs; licensed to Microsoft, became MS-DOS/PC-DOS. Also known by its working title QDOS.)
- MS-DOS (Microsoft's now abandoned DOS variant)
- PC-DOS (IBM's DOS variant)
- RDOS (Data General Corp)
- TurboDOS (Software 2000, Inc.)
- SuperDOS—an MS-DOS clone with full NTFS and USB support, based on FreeDOS. Summary of features
- Windows/386
- Windows 3.0
- Windows 3.1
- DESQview+ QEMM 386 multi-tasking user interface for DOS
- DESQView/X (X-windowing GUI for DOS)

NETWORK OPERATING SYSTEMS

- Cambridge Ring
- CSIRONET by (CSIRO)
- CTOS (Convergent Technologies, later acquired by Unisys)
- Data ONTAP by NetApp
- SAN-OS by Cisco

- EOS by McDATA
- Fabric OS by Brocade
- NetWare (networking OS by Novell)
- NOS (developed by CDC for use in their Cyber line of supercomputers)
- Novell Open Enterprise Server (Open Source networking OS by Novell. Can incorporate either SUSE Linux or Novell NetWare as its kernel).
- OliOS
- Plan 9 (distributed OS developed at Bell Labs, based on UNIX design principles but not functionally identical)
- Inferno (distributed OS derived from Plan 9, originally from Bell Labs)
- Plan B (distributed OS derived from Plan 9 and Off++ microkernel)
- TurboDOS (Software 2000, Inc.)

Web operating systems

- DesktopTwo
- G.ho.st
- YouOS
- Browser OS
- eyeOS

Personal digital assistants (PDAs)

- Inferno (distributed OS originally from Bell Labs)
- Palm OS from Palm Inc; now spun off as PalmSource
- EPOC originally from Psion (UK), now from Symbian, preferred name now is Symbian OS
- Windows CE, from Microsoft
- Pocket PC from Microsoft, a variant of Windows CE.
- Windows Mobile from Microsoft, a variant of Windows CE.
- Linux on Sharp Zaurus and Ipaq

- DOS on Poqet PC
- Newton OS on Apple Newton Messagepad
- VT-OS for the Vtech Helio
- Internet Tablet OS based on Debian Linux and deployed on Nokia's Nokia 770, N800 and N810 Internet Tablets.

Music players

- ipodlinux
- Pixo
- RockBox

Smartphones

- BlackBerry OS
- Embedded Linux
- Android
- Openmoko Linux
- Mobilinux
- MontaVista
- MotoMagx
- Qtopia
- LiMo Platform
- iPhone OS
- JavaFX Mobile
- Palm OS
- Symbian OS
- Windows CE
- Windows Mobile

COLOPHON

Webster's Revised Unabridged, copyright 1996, 1998, MICRA, Inc.:

\Col"o*phon\ (k[o^]l"[-o]*f[o^]n), n. [L. colophon *finishing stroke*, Gr. kolofw`n; cf. L. culmen top, collis hill. Cf. Holm.] *An inscription, monogram, or cipher, containing the place and date of publication, printer's name, etc., formerly placed on the last page of a book.*

And, according to the American Heritage dictionary, Colophon was an ancient Greek city of Asia Minor northwest of Ephesus, which was famous for its cavalry.

The current usage of the colophon in the publishing industry is to describe the fonts used in the book. But I think this is wrong-headed: the last page in the book is too important to be devoted to technical minutiae. So I always like to find a substantive finishing stroke for each book that I publish. In this case, what could be more appropriate than **the blue screen of death?**

—Fred Zimmerman, Nimble Books LLC,
Ann Arbor, Michigan, USA, 2008

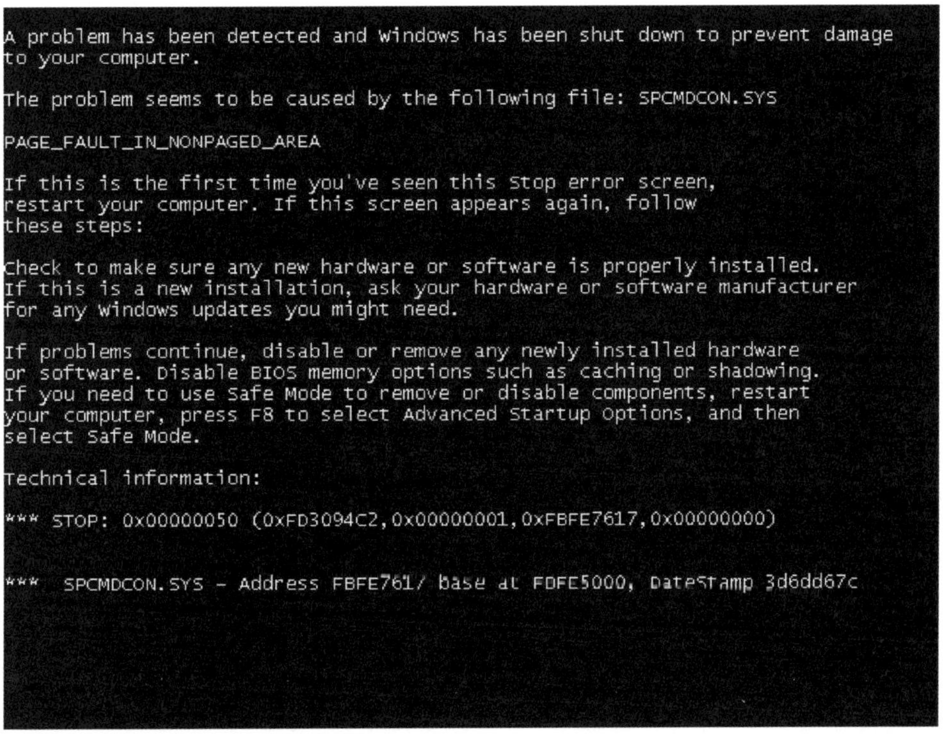

Figure 8. When a Windows operating system fails: the blue screen of death.[27]

---

[27] http://en.wikipedia.org/wiki/Image:Windows_XP_BSOD.png

Made in the USA
Middletown, DE
17 May 2021